# the very best of
# AFRICAN WILDLIFE

Gerald Hinde

DEDICATION
For my family and friends.

*And the Lord God took man and put him*
*in the Garden of Eden to tend and guard*
*and keep it.*

GENESIS 2:15.

# INTRODUCTION

As a child I spent annual holidays in the Kruger National Park and this nurtured my love for animals and wild places. On my first visit, I had a glimpse of a leopard as it leaped from a rock onto an impala ram, stunning it, before rushing off at the approach of our car. We watched as the impala slowly regained consciousness and ran off, with the leopard in hot pursuit. This was the start of a long fascination with Africa, one that would eventually take me from the boardroom to the bushveld.

Over the years, my interest in both wildlife and photography became an obsession. However, the time I could devote to pursuing my passion was limited by my commitment to our family motor business. It was only after the company was sold, in 1989, that I finally had the opportunity to work full time on a project I had begun a few years previously, namely my book *Leopard*.

The sights, sounds and smells of the bush took on a new dimension as I shed my business suits and smart sedans for khakis and a four-by-four. The elation I felt as I headed for Mala Mala Game Reserve to continue my leopard studies and photography was heightened by the realization that, at last, I had the freedom to explore Africa's great wild places, fulfilling the dream that had dominated my mind for so many years.

"*To protect our children's future we must work together in caring for our fragile earth and, simply, stop taking it for granted. It is, after all, the only one we've got.*"

After giving birth, a zebra mother keeps all other zebras at bay for several days, giving her foal time to imprint on her and learn the unique pattern of stripes that will enable them to locate each other in the herd at any time.

A Goliath Heron will wait for long periods beside water for an opportune moment to strike at a passing fish. In the five hours that I waited alongside the Sabie River for this photograph, the heron caught two tiger fish.

Giant Kingfishers, the largest of the family in southern Africa, feed primarily on small fish that they catch by diving into rivers and other bodies of water.

ABOVE After a rain shower, a lioness licks her cub's coat. This serves the dual purpose of bonding and drying the cub. Lionesses form strong bonds within a pride and cub survival is highest when reproduction is synchronized in order to facilitate communal care.

RIGHT Primates are the most socially active mammals and their resemblance to humans makes them a treat to watch as they go about their daily activities.

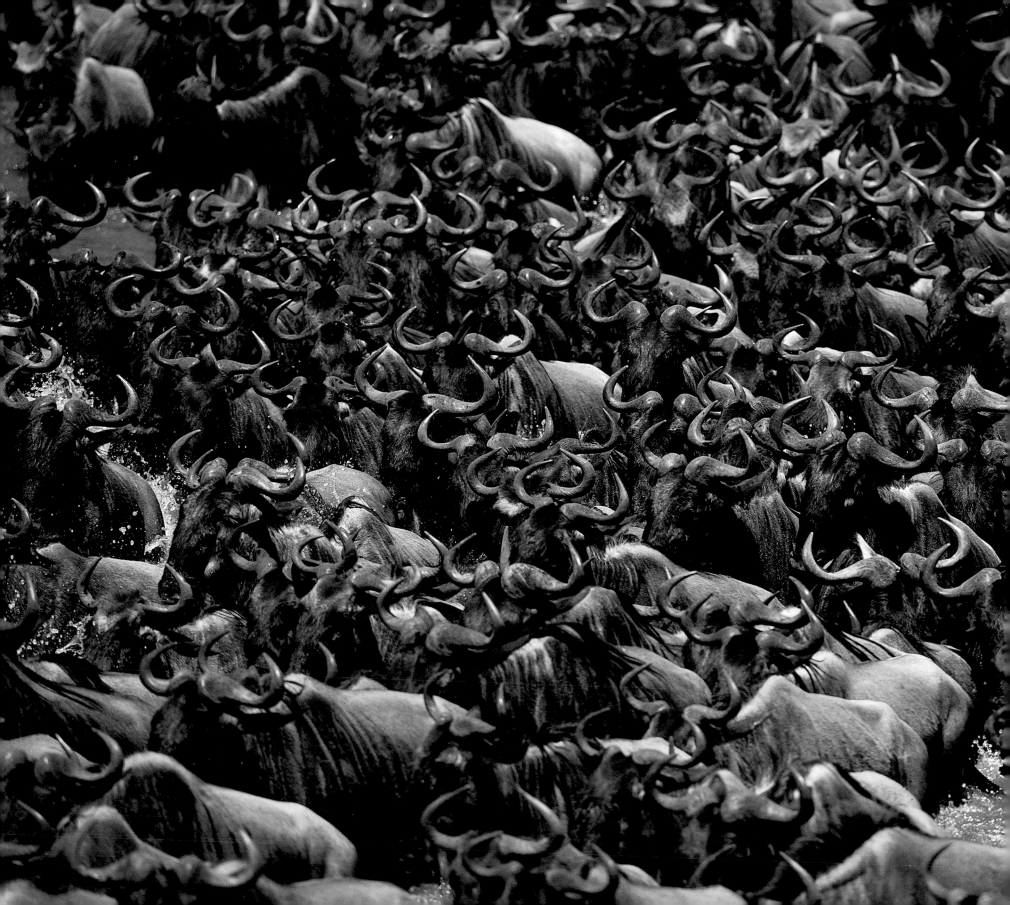

The migration of wildebeest between Kenya and Tanzania is one of the greatest wildlife shows on earth. The wildebeest population grew from 250,000 animals in 1961 to 1.3 million in 1977 and, contrary to expectations, the numbers have remained constant since then, being held in check by the availability of dry season grazing and natural culling by predators.

Wildebeest numbers build up for hours on the banks of the Mara River until, finally, one brave animal takes the plunge, to be followed by thousands of others as the epic river crossing begins. Forty-five minutes later, it can all be over.

Leopards have a very wide habitat tolerance, ranging from rainforests to deserts. Although they have a lifespan of up to 20 years in captivity, it is typically shorter in the wild, around 12 years. I was fortunate to study one particular leopard in Sabie Sands Game Reserve in South Africa, that lived for 17 years, the oldest recorded lifespan of a leopard in the wild.

ABOVE AND LEFT Hippos like to submerge themselves in water because their thin skin is vulnerable to sunburn. Contrary to belief, they are not expert swimmers, preferring to float and push themselves along with strides on the river bottom. On land, they are extremely agile and can run at 30 kilometres per hour. Their aggressive nature and powerful jaws pose a threat to humans.

OPPOSITE There are relatively few black rhino left in Kenya but, during my trips to the Masai Mara National Park, I regularly saw this cow and calf. Rhino are selective feeders, and vulnerable to malnutrition because of their sedentary habits.

TOP LEFT Spiders, a class of arthropod, have four pairs of legs and a body in two principal sections; the cephalothorax (combined head and thorax) and the abdomen. Sixty-three spider families occur in Southern Africa.

TOP RIGHT AND BOTTOM LEFT Skimmer dragonflies belong to the Libellulidae, the most common family of dragonflies. They are strong, fast fliers that feed on other species of insect.

BOTTOM RIGHT Found throughout southern Africa, the African Monarch is mimicked by some other butterfly species because it contains toxins that make it unpalatable to predators, which then leave it alone.

ABOVE AND TOP LEFT Sunbirds' tiny size allows them to move around bunches of flowers, sipping nectar from each blossom and obligingly spreading pollen from plant to plant (above: Marico Sunbird; top left: Variable (Yellowbellied) Sunbird).

LEFT Little Bee-eaters perch on twigs, grass or reed stems from where they hawk flying insects. They have a slender, delicate beak which is used to pluck bees and other insects from the air.

A Masai giraffe and her calves feed as rain clouds gather over the Olololo Escarpment in Kenya. A baby giraffe spends most of its first few weeks of life lying in hiding; the mother may move some distance away, leaving her calf vulnerable to attack by predators. Calf mortality rate can be as high as 70 per cent.

Elephants live in family groups consisting of one adult female and her offspring or a number of closely related females with their offspring. These family groups may coalesce to form larger herds. The society is matriarchal and, on reaching puberty, young males leave the herd of their own accord.

PREVIOUS PAGES There are few more fascinating sights than watching a herd of elephants come down to drink. With no warning, they appear silently out of the bush, moving at a rapid but fluid pace. After quenching their thirst and sluicing water over themselves, they disappear as suddenly as they appeared. Elephants drink about 160 litres of water a day, often chasing off other species which may be competing for limited supplies of water.

The elephant's trunk is an organ of extreme dexterity, capable of serving the huge beast like a pair of hands. It plucks bundles of grass, digs for water, removes the foetal membranes from the newly born, assists young calves out of difficulty and powders the body with dust. The tip of the trunk is extremely sensitive and is often held high to catch air-borne scents.

Misty mornings usually signal the onset of autumn as the early morning temperatures drop, causing condensation that forms the tiny droplets of water that float through the air. A waterbuck (top) and impala (above) in Sabi Sands, South Africa, stand out against the hazy background.

RIGHT Buffalo and cattle egrets in the Okavango Delta, Botswana.

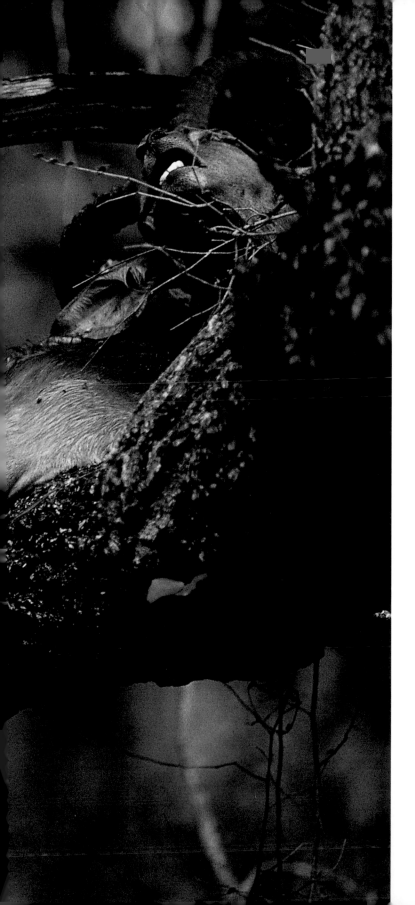

ABOVE Leopards are expert climbers from a young age, which helps them to escape from the threat of being killed by lions and other predatory animals.

LEFT In areas where competing predators occur, leopards are inclined to drag their kills into trees, but where there is no threat from other large predators, such as lions or hyenas, they will eat their catch on the ground.

ABOVE Pelicans catch fish by gathering in flocks and driving the shoals into shallow water where they are scooped up in the pelicans' bill-pouches.

OPPOSITE In Namibia, Red-footed Falcons and Tawny Eagles hunt the doves and quelea that visit the waterholes in great numbers in the early mornings. Falcons often knock down their prey when attempting to seize it with their talons in mid-flight. The fallen birds then become food for Secretarybirds, but not without fierce competition from the eagles.

A group of Burchell's zebra is startled while drinking – these animals are most vulnerable to sudden attack by predators at waterholes. The richly striped coat helps to camouflage the animal; confusing predators when they try to single out an individual in the group.

Burchell's zebras are gregarious and live in family groups that consist of four to eight animals. They are highly dependent on water, and family groups assemble in herds of many hundreds at waterholes during the dry season. This often causes conflict as the stallions protect their groups, aggressively biting, kicking and rearing to ward-off competitors.

Impala normally drop their lambs in early summer, but the time of year can vary from one area to another. All the births in a herd occur within a limited period, thereby providing predators with more than enough prey and ensuring a good survival rate for the lambs. Crèche groups are formed within the herd and lambs may become temporarily separated from their mothers at times.

Impala are found in varied habitats, but prefer light woodland or bushveld with water nearby, as they need to drink daily. Although they are principally browsers, eating leaves and fine twigs from a wide range of plants, they also graze, particularly in the wet season when fresh grass is abundant.

OPPOSITE The Southern White-faced Scops-Owl is found throughout southern Africa, occurring in riverine forests, woodlands and savanna. To enable them to hunt at night, owls have huge eyes that let in enough light to see clearly in one-tenth of the light that a human would require.

TOP Colourful Lilac-breasted Rollers perform noisy courtship rituals which include flying upwards in a steep climb, followed by an aerial roll.

LEFT The European Roller is similar in appearance to its southern African Lilac-breasted counterpart, but has a turquoise-blue head and underparts and is more seasonal in occurrence.

Cheetah cubs are particularly playful and acrobatic, chasing and jumping on each other's rump. This serves as a training school for hunting as it teaches them how to down prey. Cheetahs are most active in the early morning and late afternoon, when most hunting takes place. They can achieve an incredible top speed of 112 kilometres per hour but can maintain this only for about 300 metres before rising blood temperature and oxygen deficit force them to quit the chase.

During the dry months, the waterholes in Etosha Game Reserve in Namibia teem with a variety of game, including giraffe, zebra and gemsbok. Etosha pan was once a vast inland sea that gradually dried as climatic changes and topographic movement caused the Kunene River that fed it to change its course and flow into the Atlantic Ocean. The surrounding sweetveld savanna plains sustain vast herds of zebra, wildebeest, antelope, elephant and many other mammals, reptiles and birds.

Wild dog pups are important members of the pack and all the adults participate in their rearing by guarding and cleaning them and feeding them on regurgitated meat. However, despite the attention and care they receive, the mortality rate of pups is high due to disease and predation by lions and hyenas. The pups are born in disused antbear or other holes, and leave the den to run with the pack at about ten weeks.

Hippos live in herds that typically number 10 to 15 but can consist of 50 or more in areas of high density. Non-breeding males are tolerated in other males' territories as long as they behave themselves, but any challenges are met with such savage attacks from the territorial bull that the challenger may choose to live alone or join a bachelor herd.

OPPOSITE  In northern Botswana, early one winter morning, these wild dogs killed an impala ewe and the ensuing noise and commotion attracted three hungry hyenas. The wild dogs outnumbered the hyenas on this occasion and were victorious in chasing them off, but not before an impala limb or two had been stolen.

TOP  Lions tend to dislike water but in wetland conditions they adapt extremely well. These lion cubs were part of a pride that fortuitously found an elephant carcass in Botswana's Khwai River. Being used to frequent river crossings, they hardly hesitated before entering the water.

LEFT AND TOP The African bullfrog (left) has a wide sub-tropical distribution and differs from its cousin, the giant bullfrog, in that it has dark spots on a green-brown back extending to the face. The giant bullfrog (top) is olive-green with no facial markings. Like all mating messages, those sung by frogs proclaim just what type of animal they are and therefore what kind of mate they are seeking. During the mating season, there is much aggression amongst the males.

ABOVE Tree agamas, which feed mainly on insects, are covered in spiny scales. The breeding male develops a bright ultramarine head that he nods in display as he clambers on tree trunks; the females lay up to 14 eggs in a hole in the ground.

OPPOSITE These tree squirrels, sunning themselves early on a winter morning, are using a perch formed by elephants debarking a tree. At the slightest sign of danger, the squirrels will sit on a branch, twitching their feathery tails and chattering noisily, apparently scolding the intruder.

TOP Bat-eared foxes can be identified by their enormous ears and 'bandit's mask'. They are most active at night, but may emerge from their dens in the late afternoon when the heat of the day has abated.

OVERLEAF Black rhinos are notorious for their volatile tempers and, with their lethal horn and suprising speed, can be dangerous. They operate on the principle that the best defence is attack and will readily charge any presumed threat. Formerly widespread throughout southern Africa, black rhinos are now endangered everywhere. Outside of national parks and reserves, pockets still exist in protected areas in Damaraland and Kaokoland in Namibia.

Cattle egrets and buffalo herds always occur together; and I have frequently followed the flight of the egret flocks in the early mornings to locate the buffalo herd. As the herd moves, it flushes insects from the grass, to be eaten by the egrets.

Early summer is a time of blinding heat in Etosha, with the waterholes shrinking into brackish, tepid puddles. October is a time of 'short rains', when the billowing cloud masses offer hope and the possibility of showers. A giraffe takes a drink as the sky darkens, and a herd of wildebeest (opposite) wait as dark clouds roll over the parched landscape. Minutes later, bolts of lightning ignited the skies, dust devils swirled, and the rain came down in torrents, bringing welcome relief.

ABOVE Springbok occur mainly in arid regions and open grasslands, where they feed on semi-desert scrub or, after the occasional rains, on fresh grass. They can go for long periods without water, obtaining moisture from the succulent karroid vegetation.

RIGHT Buffalo and zebra head towards water for a late afternoon drink. Buffalo herds seek water in the early morning and again in the evening to quench their thirst. Bulls often indulge in mud wallowing, but the cows seldom do this.

Lions form strong social bonds within a pride and all the females will care for the cubs, even suckling those of other lionesses. Males occasionally pay attention to their cubs but their primary responsibility is to protect the pride from other males.

Male lions are somewhat bigger than the females and their shaggy manes make them appear even larger. These characteristics are disadvantageous when it comes to hunting so, for the most part, males let the lionesses do the work; once the females have killed, the males take over the spoils.

The dusty, late winter conditions create an eerie scene early one September morning, as wildebeest make their way towards a waterhole in the Nossob River in South Africa's Kgalagadi Transfrontier National Park.

Bat-eared foxes live in pairs or family groups, grooming each other and playing together when not foraging for food. This includes termites and other invertebrate prey, as well as mice, birds, lizards and dung beetles. They lick up harvester termites, and can sometimes be seen making spectacular leaps into the air to catch grasshoppers and flying termites.

When something as large as a giraffe dies of natural causes, scavengers make the most of it, gorging themselves and chasing off weaker competitors. Vultures and hyenas generally share the spoils, but the hyenas are not always tolerant of the vultures, frequently chasing them off the carcass.

Black-backed jackal mothers devote up to 90 per cent of their time to suckling their young during their first month. The average sized litter is three to four pups, but this female in Namibia had an unusually large litter of six, which she concealed in burrows around a culvert under the road. Both male and female jackals take part in rearing and feeding the young and often have helpers that assist with rearing the pups to independence, at around 14 weeks. The helpers are usually pups from the previous year's litter that have not as yet moved on.

Leopard cubs are weaned by the age of three months and at about four months start to accompany their mother on hunts. They stay with her until they are about 20 months old but affectionate reunions may take place long after independence.

This female leopard (right) travelled a long distance to fetch her 19-month-old cub and lead it back to an impala kill; along the way she periodically played with the youngster, but at times she lost patience and showed aggression towards it (opposite).

Alien plants are a problem in many wildlife areas and water lettuce (*Pistia striatiodes*) has completely covered some dams in parts of South Africa's Kruger National Park and has to be regulated annually. Hippo (left) and Purple Heron (above) are just a few of the creatures that happily co-exist with this alien weed.

ABOVE Southern Ground-Hornbills are usually found in family groups walking in loose array in search of food. Concentrated mainly in national parks and wildlife reserves, they are largely carnivorous, their diet including reptiles, lizards, mice, and frogs.

OPPOSITE A Marabou Stork, fishing at the end of the Khwai River in Botswana during a particularly dry season, makes a tasty meal of a frog. These large storks are generally found in flocks and often congregate alongside vultures at kills.

Buffalo herds have clearly defined home ranges, which show little overlap with neighbouring herds. The size of the home range depends on the availability of food and water, with those living in wetter areas, such as Botswana's Okavango Delta, having smaller home ranges than those in drier areas, where there is less grazing.

OPPOSITE Water that is rich in salts accumulates into soda lakes where bacteria and algae breed, attracting huge flocks of flamingos that feed off the algae. The Greater Flamingo sieves mud through its bill and can take in a variety of foods, but the Lesser Flamingo (shown here) feeds on nothing but floating algae, filtering it through a series of plates in the bill.

ABOVE As you travel into the heart of the Kalahari, the rainfall diminishes, the thorn bushes become more scattered and the grass thins. There are no great herds of antelope here and animal tracks are rare. Some animals, like these gemsbok, have developed techniques for survival in these arid areas.

Elephants have an extremely thick skin, which is normally grey in colour. However, due to their habit of dust bathing and mud wallowing, the animals tend to assume the colour of the soil in the area. The huge skull has a honeycomb structure that reduces its weight, and six sets of molars, which progressively replace each other until the last tooth has worn out at about 60 years of age. When an elephant is unable to chew its food properly, it slowly dies from starvation.

Like most cats, the leopard (left) and serval (above) display an impressive set of teeth. Teeth play an important part in classifying carnivores. In cats, the carnassials, or cheek teeth, are used for slicing off chunks of meat while the longer canines puncture flesh and grip prey.

ABOVE Once widespread throughout the savanna grasslands, servals are under threat from loss of habitat. When hunting, they may adopt a sit-and-wait technique, using their exceptional hearing to detect rodents and small mammals.

During the dry season in Namibia, Red-billed Quelea flock to waterholes by the thousand. These widespread, nomadic birds occur in large gatherings when not breeding. In flight, the flocks can resemble columns of smoke as they move and change direction rapidly.

RIGHT Clouds gathering over Botswana's Savuti Reserve promise rain. The 'short rains', in November, are followed by heavier, more sustained, rainfall in February and March.

BELOW Lion cubs intersperse rest time with periods of active play.

Warthogs live in family groups, or sounders, using either self-dug or disused ant bear or porcupine burrows in which to sleep safely. Warthog piglets can be comical, running around in tight circles in stop-start dashes, butting each other or losing their footing.

ABOVE AND OPPOSITE Colourful Lilac-breasted Rollers, with lilac throat and breast and blue belly, perch conspicuously on the tops of bushes to search for food, before diving to the ground to catch insects, scorpions, centipedes, snails, frogs and lizards.

ABOVE A high-casqued chameleon, found in Kenya, is almost camouflaged amongst a bed of fallen leaves.

OPPOSITE A flap-neck chameleon relishes the warmth of a sunny day in Namibia. Chameleons are highly adapted lizards, notable for their unusually long tongues. Rings of muscle are contracted to shoot the tongue out at high speed, while the sticky club-like tip ensnares their insect prey.

Puku (above) and lechwe (opposite) are similar-looking antelopes, the slightly smaller puku having a thicker neck and shorter, less sweeping curved horns (in the males). Puku are widely distributed in Central Africa, but in southern Africa they occur only along the banks of Botswana's Chobe River.

Water-loving lechwe seldom move far from permanent water. They characteristically occur on the floodplains of Namibia's Caprivi Strip, as well as in the Okavango Delta and Linyati Swamps in northern Botswana. Their elongated hooves spread sideways on soft ground, helping them to move through the muddy landscape.

OPPOSITE At dusk these African Sacred Ibis fly in formation toward their breeding colonies at the junction of the Chobe and Zambezi rivers, where Botswana, Zambia, Zimbabwe and Namibia meet. The common border is the strategic location of the remote, but rapidy expanding village of Kasane. As there are no fences between the village and the neighbouring Chobe National Park, game can frequently be found wandering down the road or through the camp sites.

A common resident, the African Sacred Ibis occurs widely throughout southern Africa, frequenting natural wetlands, man-made vleis and estuaries, as well as grasslands and croplands.

RIGHT Saddle-billed Storks are large, strikingly colourful birds; the females are distinguished from the males by their large yellow eyes and the lack of the yellow pendant wattle at the base of the bill. Here, a female Saddle-billed Stork catches a catfish in the muddy waters at the end of the Khwai River, Moremi Game Reserve, Botswana.

Lappet-faced Vultures (above) are powerfully built and dominate all other vultures at food sources. After spending the night perched on low thorn trees, vultures wait for the morning sun to heat the land. As the land warms up, bare patches of earth or rock reflect heat, creating a rising column of air known as a thermal. Vultures then laboriously make their way to the base of the thermal, catching the air beneath their wings and spiralling upwards within the narrow column of rising air. Once at the top of the thermal, they wheel effortlessly around scanning the plains for carrion or fresh kills.

LEFT White-backed Vultures squabble noisily in order to get a share of the spoils.

OPPOSITE After a wildebeest died on the open plains of the Masai Mara, Kenya, several vultures flew in to clean up the carcass. A Lappet-faced Vulture (right) confronts two White-backed Vultures (left).

Vervet monkeys are equally at home on the ground and in trees. They are primarily vegetarians, eating a variety of fruits, pods, seeds, shoots and leaves. During the day the youngsters are very active, spending their time playing and chasing each other.

LEFT  Protected by the towering lava walls of a former volcanic crater, and rimmed by a cloud-covered rainforest that feeds water to the plains below, Tanzania's Ngorongoro Crater supports an array of wildlife. Vast herds of resident wildebeest wander across the plains in search of seasonal grazing.

BELOW  Mountain tortoises are widely found throughout the region, particularly in the drier areas, such as here, in the Kgalagadi Transfrontier Park, South Africa and Botswana. Clutches of six to fifteen eggs are laid during the summer months.

The rare black-footed cat is endemic to southern Africa. Its preferred habitat is open, short grass, open savannas and semi-desert, where it can obtain some cover while it stalks its prey of small rodents, reptiles and invertebrates.

Sand cats occur in low densities, but their preferred habitat is so remote that human activity doesn't impact on them at all. They live in isolated desert areas, preying on rodents, mice, small birds and reptiles.

TOP AND RIGHT Crocodiles have an evolutionary history spanning millennia. Giant Nile crocodiles, which can reach up to five metres in length, can be found in rivers throughout southern Africa and beyond.

ABOVE Dragonflies devote the first part of their lives (as nymphs) to feeding and the second part to breeding.

ABOVE Dik-dik are miniature antelope, with adult males weighing just five kilograms. The crest of erect hair on the head of this male is an aggressive response to another male invading its territory. Dik-dik are monogamous and the male and female pair for life.

OPPOSITE Topi, which occur in scattered locations in East Africa, are related to the tsessebe. They have an elongated forehead and are built for speed with a lean body and clean limbs. Calves are kept hidden for three weeks and are weaned at six months.

OPPOSITE Egyptian Geese are territorial, and very noisy in social interactions. Males spend a lot of time warding off competing males and the fighting can become loud and aggressive.

ABOVE Plovers, such as this Long-toed Lapwing, frequent the shores of most still bodies of water. They are fearless birds and will dive-bomb intruders of all sizes, especially if their eggs or chicks in the nest are threatened. Here, an African Fish Eagle is being harrassed into moving on.

The desert elephants of Kaokoland, in northern Namibia, frequent the narrow strips of savannah adjacent to the rivers. These elephants are able to survive in adverse conditions thanks to their ability to travel long distances to find food and water, taking advantage of any rainfall within a vast area. In order to utilize every bit of available food, a desert elephant stretches to reach the pods from a tree (below).

LEFT Waterbuck are dependant on water and tend to stay close to it in order to drink on a daily basis. They are easily distinguished by an elliptical ring around the hump, and usually occur in small herds.

BELOW Klipspringers are one of the smaller antelope species. The name means 'rock jumper' in Afrikaans, aptly describing the way in which these agile antelope move from rock to rock and up steep slopes.

After the summer rains, the bush is transformed, with fresh green grasses and abundant flowers offering succulent grazing to an adult male black-faced impala (left, found in northern Namibia) and an adult male bushbuck (above). Most antelope are browsers, but will take advantage of seasonal grazing to supplement their diets.

There are many recorded cases of hippo protecting carcasses and even chasing off carnivores after they have pulled down prey near the water. This baby hippo died during a drought in Botswana, and the mother protected the carcass from crocodiles for three days before finally abandoning it to them.

White rhino have a profoundly different temperament to that of black rhino. The white rhino is larger and is generally placid and lethargic, whereas the black rhino is nervous, irritable and unpredictable. There have been reports of lions and hyenas attacking adult rhinos, with the outcome usually in favour of the rhino, although I have witnessed a pride of lions killing a two-year-old rhino calf that had become separated from its mother.

African (Jackass) Penguins are found in coastal waters along the west and south coasts of southern Africa. On land, they have a comical and ungainly gait, but once in the water, where they beat their flippers and steer with their feet, they can reach speeds of up to 36 kilometres per hour. In recent years, a number of land-based colonies have been established, making for easy viewing. The original name 'Jackass' comes from the penguins' loud, braying donkey-like call.

During June and July a cold current develops along the east coast of South Africa and, at this time, millions of sardines follow this inshore cold-water intrusion. In their wake come thousands of common dolphins (opposite) on course to intercept the sardine shoals, which they split into smaller pods, driving them to the surface. Cape Gannets (above) take this opportunity to plunge into the water and seize easy pickings.

Struik Publishers
(a division of New Holland Publishing (South Africa) (Pty) Ltd)
Cornelis Struik House, 80 McKenzie Street, Cape Town 8001
New Holland Publishing is a member of the Johnnic Publishing Group

Visit us at **www.struik.co.za** and view over 40,000 unique African
images available to purchase at Images of Africa Photo Library
**www.imagesofafrica.co.za**

First published in 2006

10 9 8 7 6 5 4 3 2 1

Copyright © in published edition, 2006: Struik Publishers
Copyright © in text, 2006: Gerald Hinde
Copyright © in photographs, 2006: Gerald Hinde

**Publishing manager:** Pippa Parker
**Managing editor:** Rod Baker
**Designer:** Janice Evans
**Editor:** Gill Gordon

Reproduction by Hirt & Carter Cape (Pty) Ltd
Printed and bound by Kyodo Printing Co Pte Ltd

ISBN 1 77007 377 9
Barcode 9781770073777

## AUTHOR'S ACKNOWLEDGEMENTS

Africa offers some of the best wildlife destinations in the world and,
over the years, I have been privileged to enjoy the hospitality of many
top southern African destinations, making my work even more
rewarding. Conservation Corporation, Wilderness Safaris, Desert and
Delta, Kwando Safaris, Mala Mala, and Okonjima Game Reserve are
a few of the private conservation destinations that have contributed
towards my success. I have purposely refrained from mentioning the
names of people who have helped me, but they know who they are,
and I wish to thank them all.

Throughout southern Africa, the national parks play a vital role
in wildlife conservation and are wonderful places to visit.

Struik Publishers has, as always, been great to work with and my
thanks go to Steve Connolly, Pippa Parker, Janice Evans, Rod Baker
and Gill Gordon in Cape Town and the Johannesburg-based sales
team of Deone Maasch, Janet Larsen, Jo-anne Lloyd, Leigh-Anne
Haveman and their colleagues.

This book is dedicated to my family and friends but I also
dedicate it to lovers of wildlife all over the world.

## PHOTOGRAPHIC ACKNOWLEDGEMENTS

All photography by Gerald Hinde with the exception of:
Gallo Images/gettyimages.com: pp 92–93, 126. Mike Myers: pp 39
(top), 51 (top), 54–55, 112–113. Richard du Toit: pp 78–79.